discovermore
Earth Science

Natural Disasters

Theia Lake

Britannica Educational Publishing
IN ASSOCIATION WITH
THE ROSEN PUBLISHING GROUP

Published in 2026 by Britannica Educational Publishing (a trademark of Encyclopædia Britannica, Inc.) in association with The Rosen Publishing Group, Inc.
2544 Clinton Street, Buffalo, NY 14224

Distributed exclusively by Rosen Publishing.
To see additional Britannica Educational Publishing titles, go to rosenpublishing.com.

Editor: Danielle Haynes
Book Design: Michael Flynn

Photo Credits: Cover Benny Marty/Shutterstock.com; (series background) Dai Yim/Shutterstock.com; p. 4 Alrandir/Shutterstock.com; p. 5 Toa55/Shutterstock.com; p. 7 FoxGrafy/Shutterstock.com; p. 9 (top) twintyre/Shutterstock.com; p. 9 (bottom) Designua/Shutterstock.com; p. 10 Maximillian cabinet/Shutterstock.com; p. 11 Fly_and_Dive/Shutterstock.com; p. 12 flashpict/Shutterstock.com; p. 13 FotoKina/Shutterstock.com; p. 15 (top) Jonah Lange/Shutterstock.com; p. 15 (bottom) https://commons.wikimedia.org/wiki/File:Tornado_Alley_Diagram.svg; p.16 Bilanol/Shutterstock.com; p. 17 SusaZoom/Shutterstock.com; p. 18 Marcus Placidus/Shutterstock.com; p. 19 ErichFend/Shutterstock.com; p. 21 (top) Real Window Creative/Shutterstock.com; p. 21 (bottom) Stanislav71/Shutterstock.com; p. 22 Alaskagirl8821/Shutterstock.com; p. 23 https://commons.wikimedia.org/wiki/File:President_Donald_J._Trump_Travels_to_California_(45931137101).jpg; p. 24 Sergii Figurnyi/Shutterstock.com; p. 25 Piyaset/Shutterstock.com; p. 26 Thomas B Lee/Shutterstock.com; p. 27 Zaltrona/Shutterstock.com; p. 28 Mongkolchon Akesin/Shutterstock.com; p. 29 Pierre Teyssot/Shutterstock.com.

Library of Congress Cataloging-in-Publication Data

Names: Lake, Theia, author.
Title: Natural disasters / Theia Lake.
Description: Buffalo, New York : Britannica Educational Publishing, [2026] | Series: Discover more. Earth science | Includes index.
Identifiers: LCCN 2024061104 | ISBN 9781641904223 (lib. bdg.) | ISBN 9781641904216 (paperback) | ISBN 9781641904230 (ebook)
Subjects: LCSH: Natural disasters--Juvenile literature. | Ecology--Juvenile literature. | Nature--Effect of human beings on--Juvenile literature.
Classification: LCC GB5019 .L35 2026 | DDC 303.48/5--dc23/eng/20250210
LC record available at https://lccn.loc.gov/2024061104

Manufactured in the United States of America

Contents

Dangerous Earth

For the most part, the Earth is a very safe place to live. Now and then, however, things get wild! A volcano erupts. An earthquake splits open the land. A flood covers fields and villages in raging water. Hurricanes blow down trees and homes.

Snowfall can be beautiful, but blizzards can bring dangerous conditions.

Natural disasters, such as this wildfire, are damaging not only to the environment, but to people and animals too.

Natural disasters affect thousands of people every year. They cause injury and death.

They destroy homes and buildings. We should never underestimate the power of nature. Let's take a look at some of the natural disasters people all over the globe experience every day.

Consider This

Look for patterns in the natural disasters you read about in this book. People have found ways to limit the damage such disasters cause. Why then do natural disasters continue to happen, year after year?

Erupting Volcanoes

Volcanoes look a little like mountains but can be much more dangerous when they erupt. A volcanic eruption happens when fountains of hot, melted rock, called lava, spurt out of a hole in Earth's surface.

When big volcanoes explode, they shoot out broken rocks along with lava. The rocks fall to the land, and smoke makes the air black.

The surface of Earth is a thin layer of hard rock called the crust. Below it is hot liquid rock called magma. A volcanic eruption begins when magma rises up through the crust. As it rises, the hot rock melts anything in its way, making a tunnel to the surface. Then the volcano erupts! Volcanoes can cause a lot of destruction but are also responsible for creating new landforms.

Consider This

What do you think the effect is when volcanoes cause eruptions under the sea? How do you think volcanic islands are created?

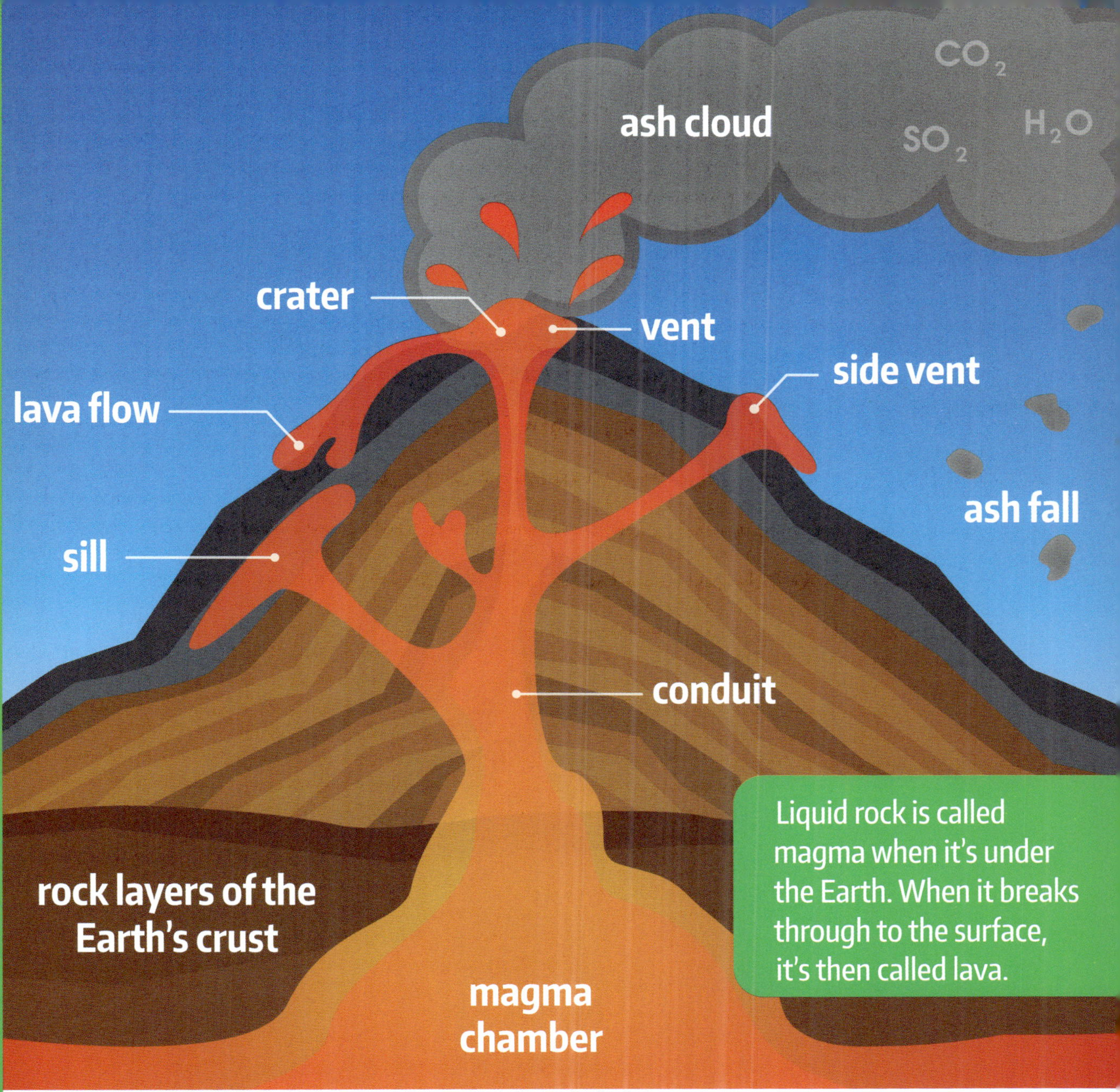

Liquid rock is called magma when it's under the Earth. When it breaks through to the surface, it's then called lava.

Shuddering Earthquakes

The geological events that cause volcanoes also cause earthquakes. Earthquakes cause more damage than any other type of natural disaster. Small earthquakes make the ground tremble. They might shake a cup off a table. Large earthquakes can bring down buildings, crack open the ground, and buckle bridges. In a big earthquake, the land can ripple up and down like waves on the sea.

Most earthquakes are caused by changes in Earth's outermost shell, or crust. The crust is made up of about a dozen rock masses called tectonic plates that are constantly moving. In different places they move apart, collide, or slide past each other. Over time, this movement causes great pressure to build up. This pressure will release, causing these plates to move along a fault, or crack in the crust.

Strong earthquakes, such as the one in Türkiye in 2023, can cause a huge amount of destruction.

compare and contrast

What happens when a volcano erupts? What happens when an earthquake occurs? How are they alike? How are they different?

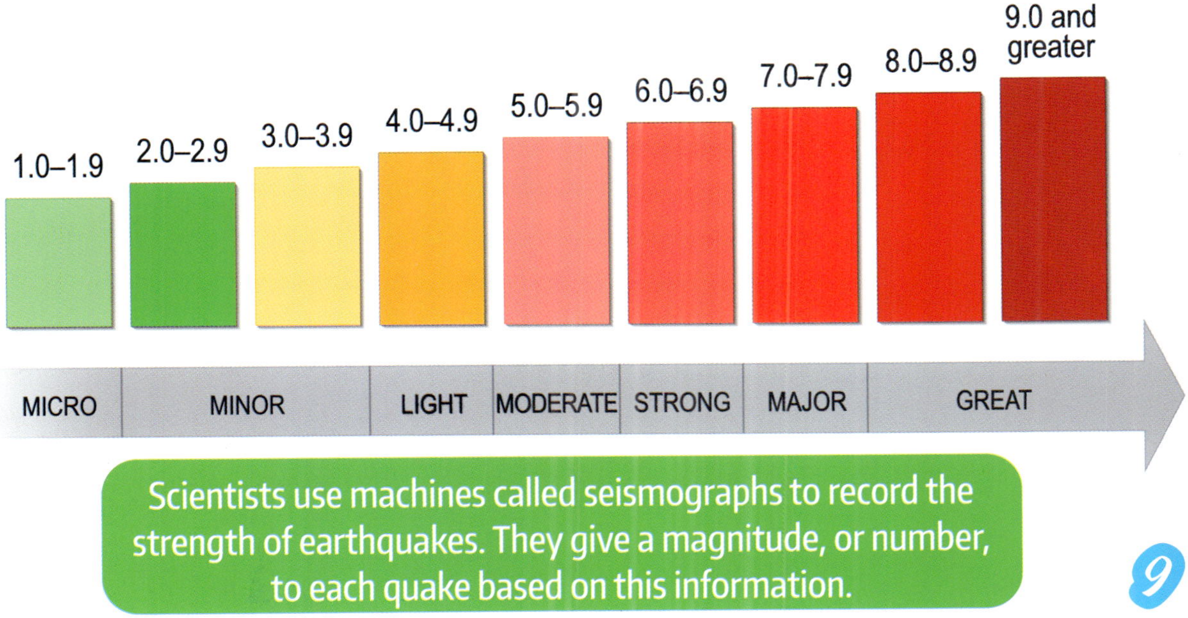

Scientists use machines called seismographs to record the strength of earthquakes. They give a magnitude, or number, to each quake based on this information.

Fast-Moving Tsunamis

Earthquakes and volcanic eruptions can also cause tsunamis. A tsunami is a huge wave that becomes bigger as it comes closer to land. Tsunamis can be 100 feet (30 m) tall. They crash onto harbors and coasts as enormous walls of water. Tsunamis can wash boats onto land and sweep away cars, buildings, and trees. They can bury fields and towns under water.

There aren't many photos of tsunamis because it's hard to know when they'll happen, and they're fast moving and very dangerous.

Tsunamis destroy everything in their paths. This damage happened after the Great East Japan Earthquake and tsunami in 2011.

Tsunamis usually happen because of an earthquake on the ocean floor. If one of Earth's plates suddenly slips below the ocean floor, this sends an enormous shudder through the seawater. The shudder creates massive waves. Underwater volcanoes can also cause tsunamis. They happen when volcanic eruptions cause large blocks of ice or rock to fall and crash into the sea. The word "tsunami" comes from the Japanese words tsu, meaning harbor, and nami, which means wave.

Consider This

How is the way that volcanoes and earthquakes cause tsunamis similar?

Blowing Hurricanes

While plate tectonics cause natural disasters such as earthquakes and volcanoes, hurricanes are a form of weather. They begin over warm seas. The warm water heats up the air just above it. When air heats up, it becomes lighter and rises. Cooler air fills the space the warm air leaves behind. This creates winds. When hot air rises quickly, the winds move faster and start to spin. When the winds spin quickly, they become a hurricane.

Seen from above, hurricanes are a beautiful swirl of clouds. Hurricanes are called typhoons in East Asia.

Hurricane winds can push seawater from the ocean onto land. This is called storm surge.

Some hurricanes stay over the ocean. They can whip up massive waves that cause floods. Others move at high speed toward land. The strong winds and the heavy rain hurricanes bring can last for days or weeks. Hurricane winds can damage buildings and **uproot** trees, while flooding leads to other problems.

WORD WISE

TO UPROOT SOMETHING IS TO PULL IT OUT OF THE GROUND.

Powerful Tornadoes

Another form of destructive weather is a tornado. This is a spinning tower of air that twists down to the ground from a thunderstorm cloud. Most tornadoes are small, last for a few seconds, and do no damage. A few are hundreds of feet wide and miles tall and can last for several hours. Like huge vacuum cleaners, they suck up everything in their way and drop it down again later.

Tornadoes start above land during a storm when warm, wet winds from one direction meet colder, drier winds coming from another direction. The warm air rises over the cold air and starts to spin. The rising, spinning air sucks in warm air from just above the ground. This warm air rises and makes the tornado higher and higher. While smaller than a hurricane, a tornado's winds are often stronger.

The part of a tornado that comes down and touches the ground is called a funnel.

compareandcontrast

How are tornadoes and hurricanes similar? How are they different?

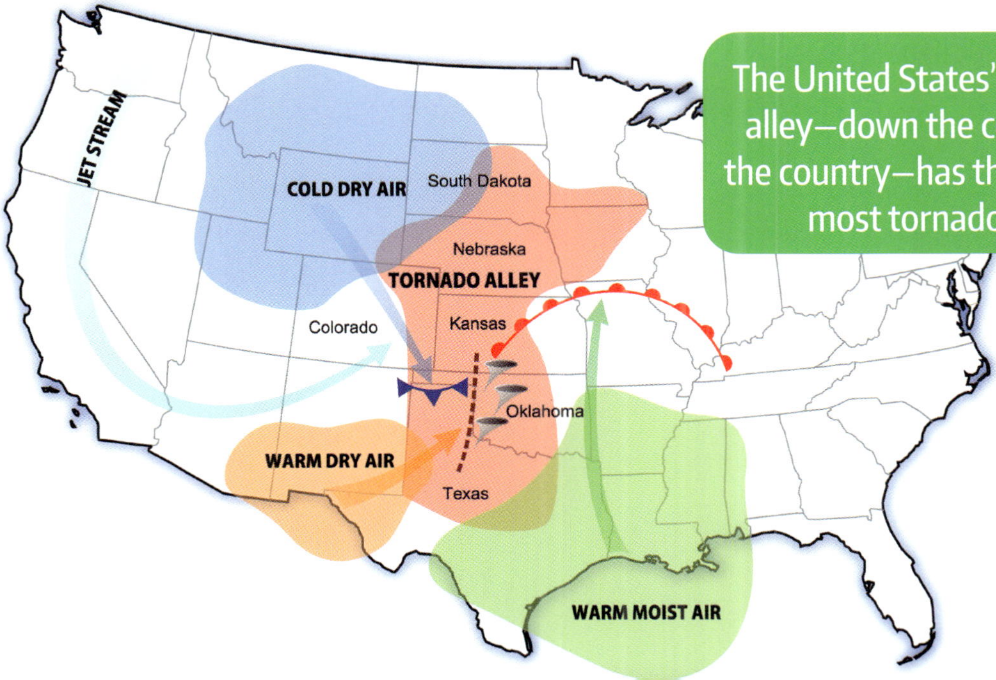

The United States' tornado alley—down the center of the country—has the world's most tornadoes.

Damaging Floods

Flooding is one event that can be linked to many of the other disasters already talked about. When water overflows onto dry land, a flood takes place. Floods can happen without any warning. Large amounts of water are very heavy, and when water moves fast, it can be extremely forceful. When water suddenly floods an area of land, it can wash away people, cars, and other objects. Floods also fill buildings with water and mud and cover fields, destroying the crops that grow there.

Flooding is more common in low-lying areas, such as Florida.

Flooded buildings will often grow mold and require lots of work to fix.

Some floods happen when a lot of rain falls in a short time. The rain fills rivers until they spill over their banks. In Asia and Africa, heavy rain falls during the **monsoon** season, causing floods. Some floods happen when tsunamis or large waves caused by hurricanes hit land. Hurricanes can cause storm surges that also flood the land.

WORD WISE

MANY PEOPLE THINK OF A MONSOON AS A DRENCHING SUMMER RAIN. HOWEVER, A MONSOON IS ACTUALLY THE WIND PATTERN THAT CAUSES SUCH RAINS.

Icy Avalanches

Avalanches can quickly ruin a fun day of skiing or snowboarding. An avalanche is a large amount of snow that moves quickly down the side of a mountain. The avalanche can knock down everything in its path.

When an avalanche hits trees, rocks, and other objects with great force, it often drags them along. When it reaches the bottom of a mountain, the avalanche can bury whole towns in thick, heavy snow.

From the bottom of a mountain, an avalanche looks like a wall of snow barreling straight at you.

Some ski slopes use avalanche control—such as this series of fences in Austria—to protect people.

Some avalanches happen after it has rained. When an area of snow gets very wet and heavy, it can suddenly slide down a slope. Avalanches can also happen when there are heavy winds or after it has been snowing heavily. Avalanches can bury people under the snow, putting them at risk of **hypothermia**.

WORD WISE

HYPOTHERMIA IS WHEN A PERSON'S BODY GETS SO COLD IT STARTS TO SHUT DOWN.

Crumbling Landslides

Landslides are like avalanches but without the snow. A landslide is a large amount of rock, soil, or mud that moves down a steep slope. Some landslides move slowly, but others can race downhill very quickly. When large amounts of rock and mud roll down a slope, they knock things down. Whole towns at the bottom of slopes can be buried in mud during a landslide.

Landslides happen when a layer of earth or rocks separates from the layer below it. Gravity pulls the loose layer downward. Mud on a slope can collapse when heavy rains make it soft and heavy. People can cause landslides when they cut down trees on a slope. Tree roots help fix the soil to a slope. Areas where trees have been cut down are more likely to have landslides.

The large Oso landslide in Washington killed 43 people after it buried a small neighborhood.

compareandcontrast

How are landslides and avalanches similar? How are they different?

Smaller landslides can cause problems for cliff-side roads. This boulder, or huge rock, is blocking cars from passing.

Scorching Wildfires

Wildfires are one of the few natural disasters that can also be caused by humans. A wildfire is an uncontrolled fire that spreads through a wide area of forest, bushes, or grasses. Some wildfires start when lightning strikes dry plants. However, people start most wildfires, usually accidentally. They can start when campfires get out of control or are not put out properly when people leave. Some fires start when people drop burning cigarettes.

In addition to flames, the smoke from a wildfire can be very dangerous to people who live nearby.

One of the most deadly wildfires in U.S. history was the 2018 Camp Fire, which destroyed nearly the entire town of Paradise, California.

After a wildfire starts, the flames spread quickly from plant to plant. A wildfire soon gets out of control. Large wildfires last for several days and destroy many plants. They also kill the animals that live among the plants. Wildfires can reach towns, causing people to evacuate, or leave.

Consider This

Why do you think there are more wildfires after a long spell of very hot, dry weather?

Heat and Droughts

One weather condition that can cause a wildfire is a drought. A drought is when a place gets little or no rain for a long period of time. Droughts happen most often in places that are already hot and dry, such as parts of the United States, Australia, and Africa. Without water, soil becomes dry and begins to crack. Plants need water to live, so they start to die when soil dries out.

A drought in the U.S. Southwest caused Lake Mead's water levels to drop sharply in recent years. Millions of people depend on this lake for drinking water.

Without enough water for crops, land dries up and cracks, killing the plants.

Drought has killed more people than any other type of natural disaster. If a drought happens in a poor country, people cannot buy water from elsewhere. When plants die, it interrupts an area's food chain, leaving animals and people without food.

Consider This

All living things need water to survive. What do plants and animals use water for?

Freezing Blizzards

A little bit of snow can be fun, but a blizzard can be dangerous. A blizzard is a powerful snowstorm that brings strong winds and lots of snow and ice. In a blizzard, winds blow at 35 miles (56 km) per hour or faster for at least three hours. Winds swirl ice and snow around in the air, causing **low visibility**. This makes it difficult for people to see where they are going.

During a blizzard, so much snow blows around, making it difficult to see far away. Think about how hard this would be to drive in.

If too much snow builds up during a blizzard, it can cause roofs to collapse or fall.

Blizzard winds blow snow into huge piles called snowdrifts. If heavy snowdrifts build up on houses, they can cause roofs to collapse. Snowdrifts also block roads and trap people in their homes, so they cannot travel to get food. Snowdrifts can be so deep that they can bury houses and trains. Snowdrifts and low visibility can make it hard for people to get home if they're outdoors, and they might get hypothermia.

WORD WISE

LOW VISIBILITY IS WHEN YOU CANNOT SEE VERY FAR AHEAD OF YOU.

Preparing for Disaster

Many of these natural disasters are becoming worse and happening more often because of **climate change**. As Earth gets hotter, there will be more fires and droughts. When ice at the poles melts, the level of the sea will rise and cause floods. We can reduce the risk of climate change by using fewer fossil fuels.

Melting ice at the Earth's poles is raising sea levels, making flooding more common in low-lying areas.

Venice, Italy, often floods because it is at sea level and has many canals. The city builds these foot bridges so people can walk around without getting wet.

People can take steps to reduce the risk of natural disasters. People in areas where tornadoes are common build safe shelters in which to hide until a tornado has passed by. People build special fences on mountain slopes to stop avalanches. Try to notice ways the people in your community try to prepare and keep safe if a natural disaster strikes your area.

WORD WISE

CLIMATE CHANGE IS THE INCREASE IN EARTH'S TEMPERATURE CAUSED PARTLY BY PEOPLE BURNING FUELS SUCH AS COAL AND OIL.

Glossary

crumble To fall apart.
crust Earth's hard, rocky outer layer.
damage The loss or harm to something or someone.
danger Something that can cause damage, injury, or loss.
eruption When a volcano explodes and hot, melted rock, called lava, spurts out of it.
fossil fuel Coal, oil, and natural gas formed millions of years ago from the remains of dead plants and animals.
freeze To become so cold that water becomes solid or turns into ice.
geological Having to do with rocks or the formation of the Earth.
gravity A pulling force that works across space.
mold A growth of damp or decaying matter that can be dangerous to eat or breathe in.
plates Giant piece of rock that make up Earth's crust. The plates float on a layer of hot, melted rock.
poles The ends of Earth. There is a North Pole and a South Pole.
scorch To burn.
shelter A safe place, for example, underground, where people can go during a tornado.
snowdrift A huge bank or pile of snow.
thunderstorm A storm with loud bangs of thunder and flashes of lightning as well as heavy rain.

For More Information

Books

Foxe, Steve. *Deadly Natural Disasters.* North Mankato, MN: Capstone Press, 2024.

Taylor, Diane C. *The Science of Natural Disasters: When Nature and Humans Collide.* River Junction, VT: Nomad Press, 2020.

Watts, Claire. *Natural Disasters.* New York, NY: DK Publishing, 2022.

Websites

Britannica Kids: Natural Disasters
www.kids.britannica.com/kids/article/natural-disasters/609706
Learn more about natural disasters on the Britannica Kids website.

National Geographic Kids: Volcano Facts
www.natgeokids.com/uk/discover/geography/physical-geography/volcano-facts
Read 17 facts about volcanoes on the National Geographic Kids website.

Index